Science Concepts SECOND SERIES

Photosynthesis

Alvin Silverstein, Virginia Silverstein,
and Laura Silverstein Nunn

Twenty-First Century Books
Minneapolis

Twenty-First Century Books
A division of Lerner Publishing Group, Inc.
241 First Avenue North
Minneapolis, MN 55401 U.S.A.

Website address: www.lernerbooks.com

Library of Congress Cataloging-in-Publication Data

Silverstein, Alvin.
Photosynthesis / by Alvin Silverstein, Virginia Silverstein, and Laura Silverstein Nunn.
p. cm. — (Science concepts)
Includes bibliographical references and index.
ISBN-13: 978–0–8225–6798–1 (lib. bdg. : alk. paper)
1. Photosynthesis—Juvenile literature. I. Silverstein, Virginia B. II. Nunn, Laura Silverstein. III. Title.
QK882.S5283 2008
572'.46—dc22 2006022566

Manufactured in the United States of America
2 3 4 5 6 7 – DP – 14 13 12 11 10 09

Contents

chapter one

Food for Life

What is the most important process that takes place on Earth? Many scientists would answer: photosynthesis. This process has made our planet livable for millions of species. It also provides the materials and the energy that Earth's creatures need to live and grow.

Food from the Sun

All living things need energy in order to live. Most of the energy that living things on Earth use comes from the Sun. Sunlight is pure energy. But many of Earth's creatures cannot use light energy directly. Green plants and a few other kinds of organisms can capture the energy in sunlight and change it into chemical energy. This chemical energy is stored as food—energy-rich compounds that can be broken down when energy is needed. Almost all other living things get their energy, either directly or indirectly, from the food made by plants.

Photosynthesis comes from two Greek words meaning "light" and "putting together." In photosynthesis, plants use sunlight energy to put together carbon dioxide from the air and water from the soil to make sugar.

During photosynthesis, oxygen is given off as a byproduct. Living things can use this oxygen "waste product" to get energy out of food. When a log burns

Sunlight is the main source of energy for most living things on Earth.

in a fireplace, oxygen from the air combines with chemicals in the log, releasing their stored energy as heat. In a car engine, the burning of gasoline releases energy that is used to run the vehicle. In a similar way, oxygen is used in the bodies of living things to "burn" food for energy. Plants and other photosynthetic organisms supply all the oxygen that is needed in our world.

Magic Trick

Plants are like magicians pulling a rabbit out of a hat—they make food out of thin air! Plants tap sunlight energy to make sugars out of carbon dioxide gas and water. The sugars are then used for energy. When combined with minerals from the soil, they are used as building materials to make plant tissues, proteins, fats, and all the other chemicals of life.

Think about It

Is it easier for plants to live without animals or for animals to live without plants?

Scientists believe that when Earth first formed, no oxygen was in the air. Primitive plants and bacteria produced the first oxygen in our atmosphere by photosynthesis. Animals appeared later,

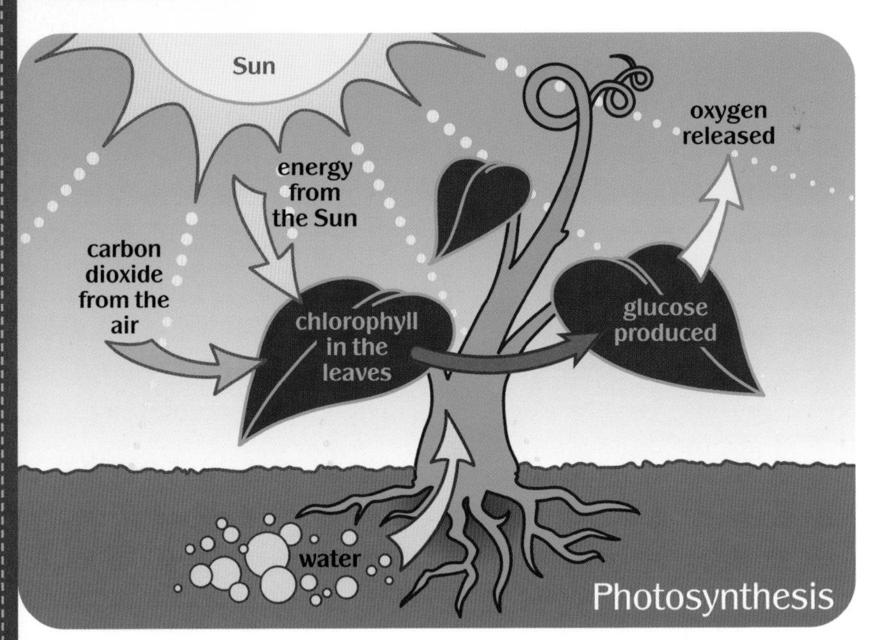

Photosynthesis is the process by which green plants make their food. The process begins when sunlight interacts with the chlorophyll in the plant's leaves and light energy is stored as chemical energy in the chloroplasts (the handy "packets" on page 7). Water and carbon dioxide also enter the leaf, and the carbon dioxide combines with water, using the energy stored in the chloroplasts, to produce sugar. The sugar is then transported to other parts of the plant. It is either stored, used right away for energy, or used to make other food substances.

when there was enough oxygen in the air to breathe.

Green Power

If photosynthesis is the most important biological process, then the substance that makes plants green, chlorophyll, could be considered the most important biological molecule. This colored chemical is found in many plant cells, particularly in the leaves, but also in the stems.

Chlorophyll, the substance that makes plants green, is found in plant leaves and also in their stems.

Chlorophyll is a catalyst. This means that it helps chemical reactions to take place, but the chlorophyll itself remains unchanged. When a particle of chlorophyll absorbs sunlight, a series of chemical reactions take place. Light energy is stored as chemical energy in handy "packets." The plant then uses the stored energy to run more chemical reactions, which produce sugar. Once the plant has formed sugar, it can make other, more complicated food substances, such as cellulose, starches, proteins, and fats.

Just the Right Amount of Light

Some plants don't grow well unless they are in full sunlight. Others grow better with less sunlight. Begonias, for example, have special cells with rounded upper surfaces that act like magnifying glasses to concentrate light. They grow well in shaded areas. In full sunlight, the magnifying cells cause the leaves to die from too much sun.

Solar Energy

The energy that reaches us from the Sun is called electromagnetic radiation. Light is only a small part of the range, or spectrum, of electromagnetic radiation, which also includes X-rays, microwaves, and gamma rays. Our eyes can see only visible light, the rainbow of colors from red to violet. Some animals, such as bees, can also see ultraviolet light (past the violet end of the rainbow). Heat-detecting goggles can pick up infrared light (past the red end of the light we can see).

Scientists argued for a long time about what light really is. It travels in waves, like the pattern of peaks and valleys on the surface of the ocean. But it also acts as if it is bundled into tiny particles, called photons. Each type of electromagnetic radiation has a different wavelength (the distance between peaks in the wave). The longer the wavelength, the smaller the amount of energy per photon. Red light has a longer wavelength than violet light. Red light contains only about half

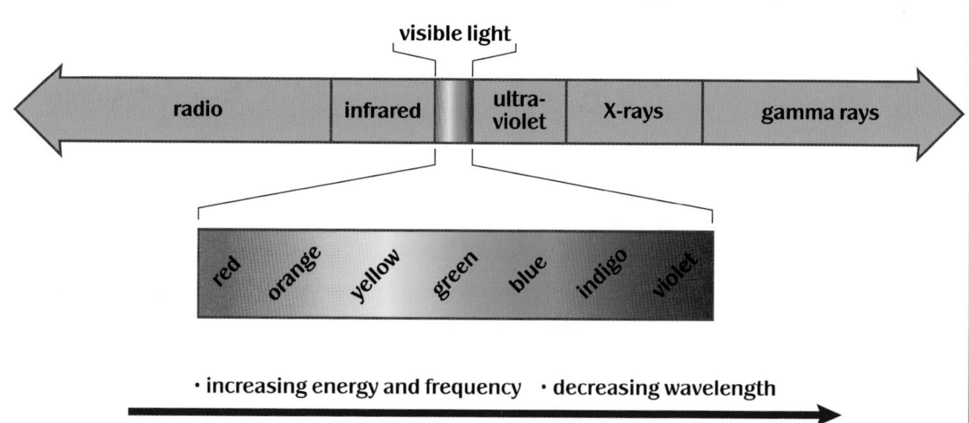

In the electromagnetic spectrum, radio waves have the longest wavelength and gamma rays have the shortest. Visible light falls in the middle.

the energy per photon. Sunlight is a mixture of all the wavelengths of light energy, from the ultraviolet to the infrared.

Making Food without Light

In the 1970s, scientists discovered communities of living organisms thriving on the deep ocean bottom, too far down for sunlight to reach. Huge red worms living in tubes, as well as new species of clams, crabs, jellyfish, and other ocean animals, were clustered around cracks, or vents, in the seafloor. This is where molten lava had oozed out from the depths of the planet, producing currents of heated water rich in dissolved minerals. What were all these animals living on? There must be some food producers to support all these consumers. But how could food producers make food without a supply of sunlight energy?

Researchers found that the food producers in these ocean vent communities are bacteria that are very different from

Tube worms that live around deep-sea vents feed off bacteria that generate food through a process called chemosynthesis. Unlike photosynthesis, chemosynthesis does not require sunlight energy for chemical reactions to take place.

those living at Earth's surface. Instead of photosynthesis, they use a process called chemosynthesis to store energy in a special compound and convert carbon dioxide to biochemicals used for food and building materials. Scientists classify these bacteria as lithotrophs (rock eaters). They get energy by oxidizing minerals such as sulfates, nitrates, and iron compounds, as well as chemicals such as hydrogen sulfide and ammonia. These are chemicals that were plentiful on the ancient Earth, before photosynthesis generated oxygen for the atmosphere.

Researchers believe that the bacteria that live by chemosynthesis are very ancient—perhaps the first living creatures that appeared on our planet. They have named them Archaebacteria, or Archaea, from the Greek word for "ancient."

In 1996 a team of British and Russian scientists

working in the southern Ural Mountains discovered the fossils of a seafloor vent community that thrived during Earth's Silurian period—438 million to 410 million years ago. The fossils included large tube worms similar to those found at modern undersea vents.

The Web of Life

Living things are linked in an intricate web of life. This web of life is also a food web, for it is made up of many food chains. Each kind of organism in a food chain uses the next type in the chain as part of its food supply.

Did You Know?

Each year 150 to 200 billion tons (136 to 181 billion metric tons) of sugar are produced through photosynthesis. Three-quarters of this food is produced in the sea.

Living things are either producers or consumers. Producers produce food. Consumers cannot make their own food. They must consume the food made by producers. If all living things were consumers, life on Earth would end when the food was all used up.

Producers make up the largest share of the biomass—the amount of living materials—on our planet. Producers are the living creatures that take energy from the Sun to make food (first in the form of sugar) through photosynthesis. These sugar makers include green plants, algae, plankton, and photosynthetic bacteria.

Primary consumers, or herbivores, get their sugar by eating plants. Herbivores include animals such as cows, deer, rabbits,

Animals such as cows are primary consumers. They get sugar by eating plants.

mice, fruit- and seed-eating birds, most insects, and the plankton eaters of the oceans.

Animals (and some plants) that feed on plant-eating animals are secondary consumers, or carnivores. Carnivores include all the members of the cat and dog family, most mammals that live in the sea, most reptiles, spiders, starfish, and even some plants, such as the Venus flytrap.

Some animals are both primary and secondary consumers. Bears, for example, eat fish and berries. Animals that eat a mixed diet are called omnivores. Humans are omnivores too.

Even at the microscopic level, organisms that can't use the Sun's energy depend on those that can. A photosynthetic bacterium, for example, is lunch for an amoeba.

Decomposers play an important role by breaking down the waste products and the dead bodies of the living creatures of all the other three groups. Decomposers include bacteria, fungi, and some animals such as earthworms, dung beetles, and maggots.

Dung beetles roll feces into balls, then lay their eggs in them. Their larvae feed on the feces. In this way, the beetles break down a waste product.

Experiment: What Do Plants Need?

1. Plant a corn seedling in each of four pots, each containing the same amount of soil.
2. Place the first pot in the light, but do not water it.
3. Water the second pot every day, but keep it in a dark closet.
4. Place the third pot in the light with a small cup of soda lime next to it. Water the pot, and cover both the pot and the cup with a large-mouth jar. (Soda lime takes carbon dioxide out of the air. The pot needs to be watered only at the beginning of the experiment because the jar keeps the water from evaporating.)
5. Place the fourth pot in the light, and water it regularly.
6. Compare the four plants after two weeks.

chapter two

Milestones along the Way

Plants have played an important part in human history. The earliest humans hunted for meat but also gathered plant fruits, roots, and leaves to eat. People had to roam from place to place to find food. They learned about what plants were good to eat and observed how plants grow. Eventually they saved some of the seeds from good food plants to grow crops the next year. People first began to grow plants for food about ten thousand years ago. This new practice changed the way early humans lived. Many of them settled down and built permanent homes.

Ancient Ideas about Plants

Aristotle (384–322 B.C.) and other ancient Greeks knew that the life processes of animals depended on the food they ate. But how did plants get *their* food? The soil was the only thing that people could see touching the plant's roots. So the ancient Greeks decided that plants must get all their food from the soil. (Air touches plants too, but people can't see or feel air.) No one came up with a better idea until about 350 years ago.

Learning how to grow new plants from seeds drastically changed the lifestyles of humans.

A Carefully Planned Experiment

A seventeenth-century Belgian physician named Jan Baptista van Helmont decided to find out for himself where plants really get the food they need to grow. First, he weighed a wooden tub filled with dry soil. Then he weighed a willow shoot and planted the willow in the tub. For five years, van Helmont added nothing but pure water to the tub. By then the willow shoot had become a small tree.

Van Helmont removed the tree and its roots, and dried the soil in the tub. The tree had gained about 165 pounds (75 kilograms) in weight, but the soil weighed only about 2 ounces (60 grams) less than it had when he started five years earlier. Obviously, the tree did not get all the materials it

Belgian physician Jan Baptista van Helmont (1579–1644) did experiments to find out where plants get the food they need to grow. His conclusions weren't exactly accurate.

needed to grow from the soil, as people had thought for the previous two thousand years! But van Helmont did not guess the real answer. Instead, he mistakenly concluded that the material that made up the bark, wood, roots, and leaves came from the water he had added over the five years.

A Closer Look at Air

The next important step in the understanding of photosynthesis occurred in the early 1770s. Joseph Priestley is the British scientist who is credited with discovering oxygen. He knew that burning a candle in a closed container used up something in the air so that the flame would eventually go out. He found that a living sprig of mint helped to "restore" the air in a chamber so that a candle could burn again. He concluded that plants helped to "cleanse and purify our atmosphere." Priestley's experiment did not work when he moved the chamber to a dark corner of the laboratory, but he did not realize the importance of light in the process. (The mint sprig needed light to keep producing oxygen, which the burning candle used to produce a flame.)

After Priestley's discovery, hospitals began putting lots of flowers in sickrooms to "cleanse" the air. A Dutch physician, Jan Ingenhousz, wanted to find out whether flowers really could help cure illnesses. In 1779, after many different tests, he determined that only the green parts of plants "cleaned" the air—and only when placed in strong light. Flowers and other non-green parts of the plant used up oxygen just like animals do. In 1796 Ingenhousz suggested that this process of photosynthesis causes carbon dioxide to split into carbon and oxygen. Then the oxygen is released as a gas.

British scientist Joseph Priestley (1733–1804) contributed to our understanding of photosynthesis by finding a connection between plants and the air in our environment.

Later, other scientists found that sugars contain carbon, hydrogen, and oxygen atoms in a ratio of one carbon molecule for each molecule of water (CH_2O). This is where the word *carbohydrate* comes from—*carbo* for "carbon" and *hydrate* for "water." (Carbohydrates are a family of chemical compounds including sugars and also starches. Starches are made up of large numbers of sugar units linked together.)

In 1804 Swiss scientist Nicholas Theodore de Saussure repeated van Helmont's experiment but carefully measured the amounts of carbon dioxide and water that were given to the plant. He showed that the carbon of plants comes from carbon dioxide and the hydrogen from water. Forty years later, a German scientist, Julius Mayer, showed that the energy of sunlight is captured in photosynthesis.

Green Magic

Scientists wondered what it was that allowed plants to make their own food while animals cannot. Was it their green color? Other living things that are similar to plants but are not green, such as mushrooms, do not photosynthesize. Even green plants photosynthesize only in the parts that are green. A tree does not photosynthesize in its roots, bark, or branches.

Two French scientists, Pierre Joseph Pelletier and Joseph Bienaime Caventou, isolated the green substance from plants in 1817 and named it chlorophyll, from Greek words meaning "green leaf." In 1883 Julius von Sachs showed that chlorophyll is not scattered all around a plant cell but is found in special structures called chloroplasts.

Chlorophyll is a complicated molecule, and it was not until 1906 that scientists began to understand more about it. German scientist Richard Willstatter found two kinds of chlorophyll. One, which he named chlorophyll a, makes up three-fourths of the

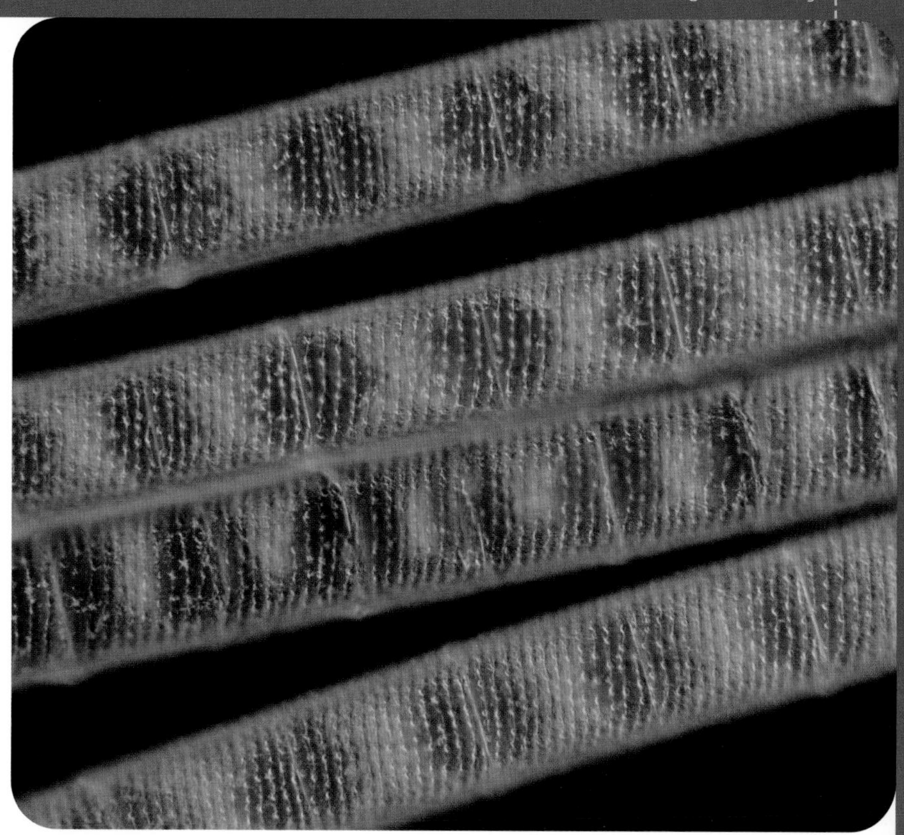

This image of green algae shows the chloroplasts, which are dark green dots within the cells.

chlorophyll in plants. The other, which he called chlorophyll b, accounts for the other one-fourth. Willstatter found that chlorophyll contained carbon, hydrogen, oxygen, and nitrogen atoms, which was not surprising. But it also contained a metal, magnesium. This was the first molecule from a living thing to be found to contain magnesium.

Willstatter received a Nobel Prize in Chemistry in 1915 for his work on chlorophyll. In 1965 a U.S. scientist, Robert Burns Woodward, won a Nobel Prize in Chemistry for figuring out the structure of the chlorophyll molecule.

Where Does the Oxygen Come From?

The overall reaction for photosynthesis was thought to be

$$CO_2 + H_2O + \text{light energy} \rightarrow (CH_2O) + O_2$$

carbon dioxide — water — carbohydrate — oxygen

For many years, scientists believed that the oxygen released came from the carbon dioxide molecule. Most accepted this idea until the 1930s, when a graduate student at Stanford University, C. B. van Niel, questioned it.

Van Niel was investigating the photosynthesis that occurred in different groups of bacteria, such as the purple sulfur bacteria. Purple sulfur bacteria create carbohydrates, but they do not release oxygen. Van Niel proposed that it was not carbon dioxide that is broken up to produce oxygen gas during photosynthesis but the water molecule.

Later, investigators used special radioactive isotopes of oxygen to show that the oxygen that is released does indeed come from the water and not the carbon dioxide. So the proper equation for photosynthesis in algae and green plants is

$$6\,CO_2 + 12\,H_2O \rightarrow C_6H_{12}O_6 + 6\,O_2 + 6\,H_2O$$

carbon dioxide — water — glucose (a sugar) — oxygen — water

Notice that water appears on *both* sides of this

equation, and the total numbers of carbon, hydrogen, and oxygen atoms on the left side are equal to the corresponding sums on the right.

The Light and Dark Sides of Photosynthesis

Photosynthesis is really a two-stage process—and light is necessary for only one of them. A British plant physiologist, F. F. Blackman, was the first to discover this. He reported in 1905 that one set of reactions depends on light and the other on temperature. The two stages are often called the "light reactions" and "dark reactions" of photosynthesis. The dark reactions use the products of the light reactions. Their names seem to mean that the light reactions occur only in the light and the dark reactions occur only in darkness. Actually, however, *both* sets of reactions take place in the light. But when a plant is in darkness, the light reactions stop and only the dark reactions continue. Light is not directly involved in the dark reactions.

Blackman found that the dark reactions go faster when the temperature is increased to 86°F (30°C), but they slow down at higher temperatures. He concluded that these reactions are controlled by enzymes—special proteins that help reactions to take place. This has been proven to be true.

In the light reactions, light energy is used to form a special compound called ATP (adenosine triphosphate). The chemical bonds of ATP can store energy and release it readily when it is needed. In the dark reactions, ATP is used to build simple sugars from carbon dioxide. The chemical energy of the ATP molecule is transferred to the bonds that hold the sugar molecule together. ATP itself loses one of its three phosphate units in the process and is converted to ADP (adenosine *di*phosphate).

In the 1940s, at the University of California, Berkeley, a young scientist named Melvin Calvin conducted a series of experiments that worked out the complicated steps in which sugars are formed. Calvin grew algae in a glass chamber shaped like a big lollipop. He bubbled radioactively labeled carbon dioxide through

Scientist Melvin Calvin holds a model of the sucrose molecule that he used in his study of sugar structure in the 1940s.

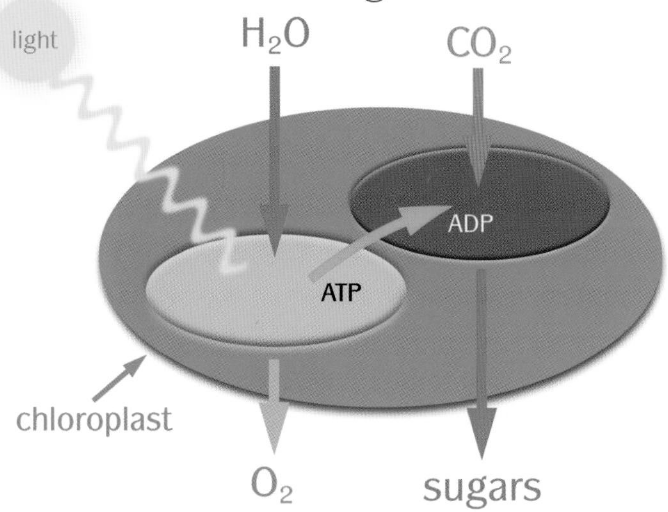

The two stages of photosynthesis are called the light reactions and dark reactions. The light reactions store light energy in a chemical form (ATP). The dark reactions use the stored energy to form sugars. Both kinds of reactions take place in the light, but only the dark reactions occur in darkness.

the suspension in the green "lollipop." This way he could follow what happened to the specially tagged carbon atoms during the reactions of photosynthesis. At various times, he dumped part of the mixture from the lollipop into a flask of boiling alcohol. That stopped the reactions immediately. Calvin analyzed the sample to see which compounds in the mixture contained radioactive carbon. Each successive sample gave him an idea of what was going on at later and later stages in the process. The light-independent, or dark, reactions of photosynthesis are called the Calvin cycle, and the scientist won a Nobel Prize for his work.

chapter three

How It Works

Plants are like busy factories. The main production areas are the leaves, where most of the work of photosynthesis takes place. But the plant "factory" also includes supply lines to bring in raw materials and transport them to the work areas. It also has energy-producing "furnaces" to power the work, storage areas for the finished products, and ways to get rid of waste products.

Roots: Absorption of Water and Minerals

Soil contains water and minerals that plants need to grow. Minerals in the soil dissolve in water, just as salt dissolves when you pour it into a cup of water. Plant roots are anchored in the soil and absorb water with its dissolved minerals.

Tiny root hairs grow out from the tips of a root. There are billions of root hairs on most plant roots. Each one is tiny, but together they add up. The root hairs absorb most of the water that plants take in. The more root hairs, the greater the surface area available to absorb water.

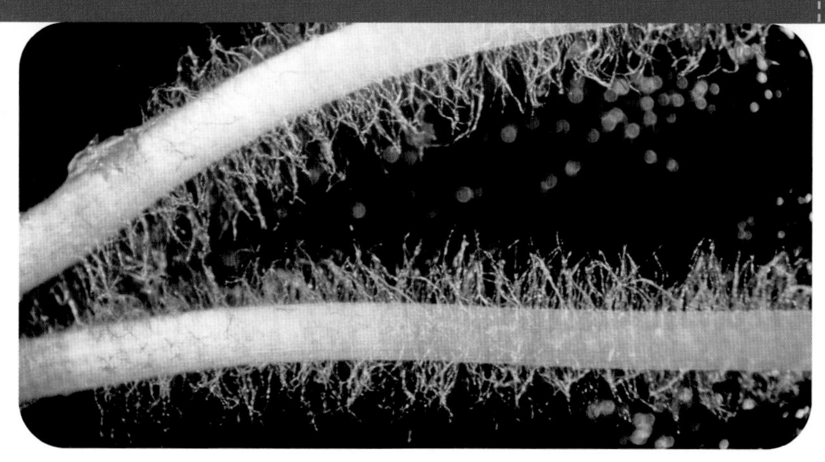

This photo of radish roots shows the tiny water-absorbing root hairs.

Stems: Transportation of Materials

Water and minerals travel from the roots up to the leaves through the plant stem. A stem contains two types of tubes for transporting materials. Tubes called xylem bring water and minerals from the roots to the leaves. Phloem tubes carry food from the leaves to all parts of the plant.

This cross section of a pine needle shows the two types of tubes that transport materials through the plant stem. Here, the xylem is colored red and the phloem is colored blue.

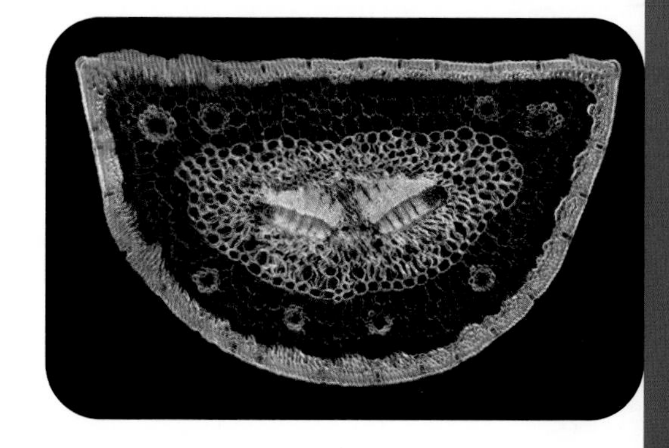

Leaves: Nature's Food Factories

Leaves are usually flat and wide. This gives them a large surface for them to absorb as much sunlight as possible. The more leaves a plant has, the more sunlight it can absorb. A full-grown

Experiment: A Look at Xylem Tubes

Put ten drops of red food coloring into a half cup of water. Cut off the bottom of a celery stalk, and place it in the colored water. Let it sit overnight, and then cut the celery in half to look at a cross section. You will see the xylem tubes have turned red.

maple tree has about 100,000 leaves, with a total surface area of 2,000 square yards (1,670 sq. meters).

Leaves are made up of many parts. The top and bottom layers protect the insides of the leaf the way a

When you look at a leaf close-up, you can see the many veins that run through it. These veins transport materials throughout the leaf as well as away from the leaf to other parts of the plant. The veins also support the leaf and help keep its shape. The biggest vein is in the center of the leaf. From there, secondary and tertiary veins branch off.

roof and floor protect a house. The top layer is clear, like a sunroof, to let in the sunshine. Beneath the outer layer is a layer of food-making cells. In another spongy layer of cells, food is stored before it is taken to the rest of the plant. The parts of the leaf are connected to the stem by veins.

Stomates: The Gateways for Gases

Carbon dioxide passes into plant leaves through tiny holes on the undersides of the leaves. Each of these special holes, called stomates, is surrounded by a pair of guard cells. When the guard cells are opened wide, gases can pass in and out of the stomates. When they are squeezed together, gases cannot enter or leave.

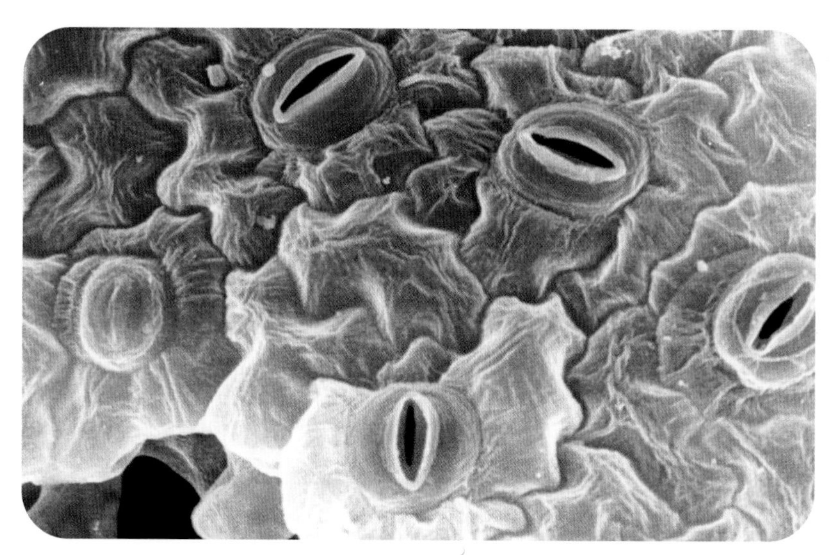

The surface of this tobacco leaf shows several opened and unopened stomates with their surrounding guard cells. The guard cells control how much water vapor and carbon dioxide pass through the stomates.

Experiment: Find the Stomates

Keep a geranium plant in a dark closet for three days. Then set the plant on a sunny windowsill. Put petroleum jelly on the top of one of the leaves and on the bottom of a different leaf. (The petroleum jelly will block up the stomates so that the leaf cannot "breathe.") Observe the geranium plant's leaves each day over the next week. (Remember to water the plant when the soil gets dry.) Compare the petroleum-coated leaves with one of the leaves that was not coated on either side. Can you tell which surface of the leaf contains the stomates?

Inside a Chloroplast

The food-making parts of plants contain chlorophyll, which makes the layer of food-making cells look green. Chlorophyll is found in parts of the cell called chloroplasts. Chloroplasts are energy-capturing organelles. (An organelle is a tiny working structure found in an individual cell.) They are both solar cells and sugar factories. They capture the energy from sunlight, and they help produce plant food—sugar.

Each chloroplast is made up of a smooth outer and inner membrane that enclose a chamber called the stroma. The stroma is filled with a thick, watery solution. Within the inner membrane is a third membrane that forms a complicated system of stacked disklike sacs, called thylakoids. These flattened sacs are interconnected by flattened channels. Each thylakoid has a space inside it, which is surrounded by the thylakoid membrane.

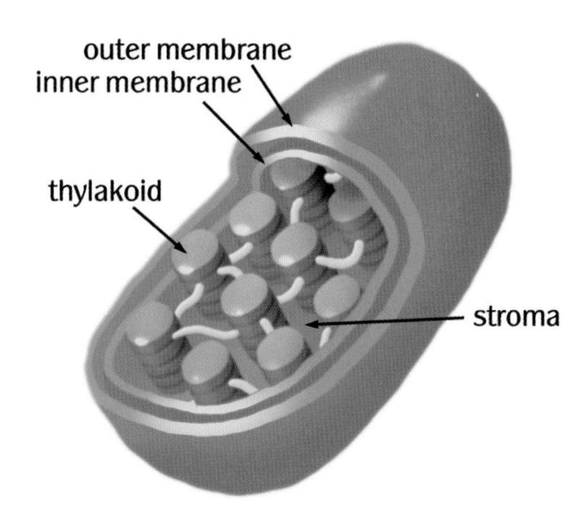

Structure of a Chloroplast

The chloroplasts in a plant cell contain chlorophyll and capture light energy. They also help produce sugar in stacks of disklike thylakoids.

Each leaf cell contains about 50 or more chloroplasts. In a square millimeter of a corn leaf, there can be 500,000 chloroplasts. The chloroplasts in all the leaves of a full-grown maple or elm tree have a surface area of about 140 square miles (360 square kilometers)! A single maple tree can produce 2 tons (2 metric tons) of food (sugar) on a sunny day.

Pigments and Light Energy

A pigment is any substance that absorbs light. If a pigment absorbs all wavelengths of light, it looks black. Some pigments absorb only certain wavelengths of light. The rest are reflected. Chlorophyll is a pigment that absorbs violet, blue, and red wavelengths. It reflects green wavelengths, and so it looks green to us. Other plant pigments involved in photosynthesis include carotenoids (red, orange, and yellow pigments) and phycobilins (red and blue pigments).

When pigments absorb light, some of their electrons gain more energy. The "excited" electrons may then give off energy as heat or light. Or the energy may be captured in a chemical bond. This is what happens in photosynthesis.

The Chlorophyll Factory

Chlorophyll and other pigments, along with some

The Other Pigments

Carotenoids and phycobilins are pigments that are also involved in capturing light energy. But this energy must be transferred to chlorophyll to be converted into chemical energy. Much more chlorophyll is in green leaves than other pigments, so the other colors are hidden. In the fall, when chlorophyll breaks down, carotenoids give some leaves their typical reddish orange fall colors.

Phycobilins are found in cyanobacteria and in the chloroplasts of red algae. Red wavelengths of light reach farther down in the sea than other wavelengths. Red algae are able to use red light for energy, so they can live in much deeper water than other seaweeds.

special proteins needed for photosynthesis to occur, are embedded in the thylakoid membranes. Chlorophyll molecules are organized into groups of 250 to 400 pigment molecules. Each group is called a photosystem. Scientists believe that two kinds of photosystems work together to produce stored energy.

All the pigment molecules in a photosystem can absorb packets of light energy called photons. But only one chlorophyll molecule in each photosystem can use light energy for photochemical reactions. This special chlorophyll molecule is called the photosystem's reaction center. The other pigment molecules are called antenna pigments. They gather light energy like a network of antennas. The antenna pigments pass the light energy they have collected along to the reaction center, which gives its high-energy electrons to a neighboring molecule (an electron acceptor).

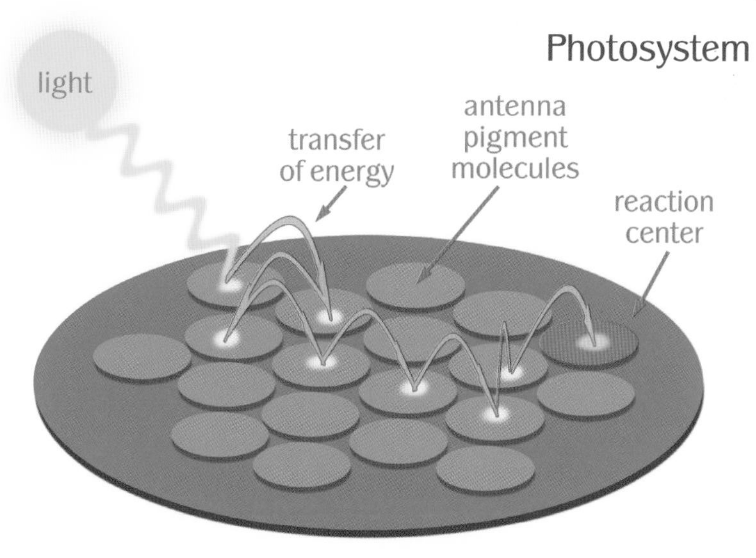

Antenna pigments gather light energy and pass it along to a chlorophyll molecule where photochemical reactions take place.

Chlorophyll can convert light energy to chemical energy only in the thylakoid membranes, working with their special proteins. When light shines on chlorophyll molecules in a test tube, the pigment molecules just absorb energy temporarily. They then release it as heat and light, without performing any useful work.

There are several kinds of chlorophyll that have somewhat different molecular structures. The blue green chlorophyll a is the most common and is found in nearly all photosynthetic organisms. Many also contain the yellow green chlorophyll b, as well. These two forms absorb light of different wavelengths. A plant that contains both can use more of the energy found in light.

Only chlorophyll a molecules can transform light energy into a chemical bond, however. When light excites a chlorophyll b molecule, the molecule must

Chlorophyll without Chloroplasts

Blue green bacteria are microscopic organisms. Some grow as single cells, some are masses of thin threads, and some form huge colonies. They contain chlorophyll, but they do not contain chloroplasts. Chlorophyll is spread throughout the cell. Many scientists believe that the chloroplasts in plant cells evolved from photosynthesizing bacteria much like these.

Tremendous Light Absorbers

Chlorophyll molecules can absorb an enormous amount of light. Almost all the red and blue light passing through a container of water will be absorbed if only a few milligrams of chlorophyll are placed on it.

transfer the energy to a chlorophyll a molecule so it can be used.

Some types of algae, such as brown algae and diatoms, have a different form, chlorophyll c, instead of chlorophyll b. Bacteria that do not produce oxygen as a waste product contain bacteriochlorophyll (in purple bacteria) or chlorobium chlorophyll (in green sulfur bacteria).

How Chloroplasts Make Sugar

Photons hit the chlorophyll molecules in leaves. Electrons are knocked into a higher-energy state. The electrons bounce along a series of chlorophyll and enzyme molecules and eventually reach small special carrier molecules in the thylakoid membrane. The electrons create a current as they flow down along this electron transport chain, and some of their energy is used to form an ATP molecule.

Meanwhile, the missing chlorophyll electrons have to be replaced or else the chlorophyll molecules can no longer be used. They are replaced by electrons from water molecules, so that the chlorophyll molecule is ready for action once again. (A special enzyme splits water molecules—H_2O—into two electrons, two hydrogen ions, and one oxygen atom.)

High-energy electrons and hydrogen ions combine with a substance called NADP (nicotinamide adenine dinucleotide phosphate) to form NADPH (NADP plus hydrogen). NADPH supplies energy to the food-making reactions of photosynthesis.

So far we have described the "light reactions," which must take place in sunlight. They occur in the thylakoid or photosynthetic membranes. They are followed by the "dark," or "light-independent reactions," of photosynthesis, which produce sugars. These dark reactions take place in a different part of the chloroplast: the stroma, the space outside of the thylakoid sac.

A team of enzymes helps finish the sugar-making process through a series of chemical reactions. Using ATP molecules for energy, carbon dioxide from the air is combined with the "hot" hydrogens on NADPH, eventually producing sugar after a series of enzyme reactions. The energy that was stored in the chemical bonds of the high-energy carriers ATP and NADPH is released and stored in the bonds of sugar molecules.

Life's Building Block

Glucose is a simple sugar, whose molecule contains six carbon atoms combined with twelve hydrogen atoms and six oxygen atoms. It is the basic building block for the other materials that make up living things. Glucose molecules are put together to form large molecules of starch, for food storage.

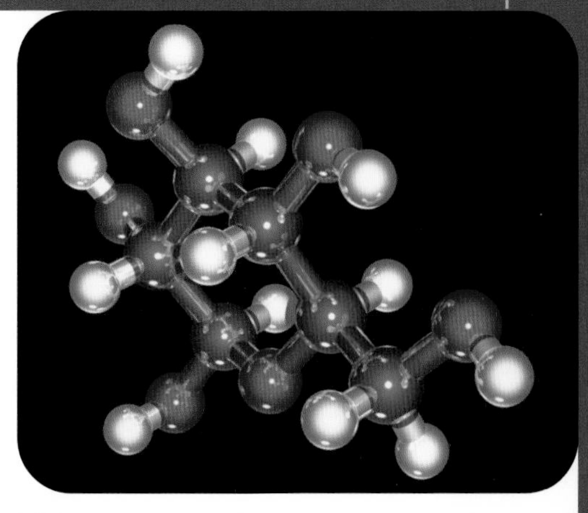

This is a model of the glucose molecule, showing atoms of carbon in blue, hydrogen in white, and oxygen in red.

Cellulose is a structural material made from even larger numbers of glucose units. Cellulose forms the outer walls of plant cells and is the main substance found in wood. Paper and cotton are nearly purely cellulose.

Glucose can be converted to fats and oils, which are more concentrated energy reserves than sugars and starches. Minerals that are absorbed from the soil are combined with glucose molecules to form other, more complicated biochemicals. Glucose can be converted to amino acids, which form the many thousands of kinds of proteins that are found in living things. Glucose can also be converted to nucleotides, the building blocks of RNA and DNA, the chemicals of heredity.

Carbon Dioxide Traps

Plants get carbon dioxide from the air. Special sugarlike molecules found in the plants act as carbon dioxide traps to hold tightly onto the carbon dioxide molecules that are absorbed from the air. Six trapped carbon dioxide molecules are needed to make one molecule of the simple sugar glucose.

If plants need sugarlike carbon dioxide traps to make sugars by photosynthesis, how do these molecules get in the plant tissues in the first place? Young plants make the biochemicals they need using the stored building blocks and energy provided by their parents.

Most green plants form seeds or spores to produce a new generation of green plants. Seeds contain a supply of starch or other food for the plant to use for energy when it is first starting out. The energy stored in the seed is used to build the roots and green shoot that grows above the ground and also to make chlorophyll, carbon dioxide traps, and the other compounds needed for photosynthesis. By the time the stored food supply is all used up, the plant is ready for photosynthesis to take over.

Life Depends on a Waste Product!

When water molecules are split into hydrogen and oxygen during photosynthesis, the hydrogen combines with carbon dioxide to form glucose molecules. Some of the oxygen is used by the plant, but the rest is released into the air. The waste product of photosynthesis in plants—oxygen—is a substance nearly all Earth's creatures cannot live without.

Why do animals need oxygen? Plants form food products using sunlight and carbon dioxide from the air. Oxygen allows animals to use the energy trapped inside this food. The oxygen that animals breathe is used to "burn" the food to release the trapped energy.

In a process called respiration, energy is released from the food materials, and carbon dioxide and water are given off. The process of respiration is the

What Happened to the Dinosaurs?

If no photosynthesis occurs, no food is produced. One theory about why the dinosaurs disappeared is that a massive comet or meteor hit Earth. The force was so great that huge clouds of dirt, dust, and other particles rose up from the ground into the atmosphere, blocking out the sunlight for a long time. (Other scientists think a large volcano spewed out thick clouds of material, with the same effect.) Without sunlight, many plants would have died, and the dinosaurs that depended on them would have died too.

opposite of the process of photosynthesis.

Two kinds of tiny organelles are responsible for the way living things handle the energy that flows through life. Chloroplasts use the energy of light to build sugar through photosynthesis. Mitochondria break down sugar in both plant and animal cells. In this process of respiration, ATP is produced. When the ATP is used by the cell, heat is released as the last stage of the flow of energy. The flow of energy through life can thus be described as:

sunlight → sugar → ATP → heat

Chloroplasts and mitochondria are similar in many ways. Both organelles are elongated and about the size of bacteria. Both carry out energy-related tasks. The two phases of respiration take place in areas similar to those in which the two

Photosynthesis vs.	Respiration
Needs sunlight	Does not need sunlight
Occurs in cells with chlorophyll	Occurs in all living cells
Uses water	Gives off water
Uses carbon dioxide	Gives off carbon dioxide
Gets energy from the Sun	Releases energy from sugar
Makes sugar	Breaks down sugar
Gives off oxygen	Uses oxygen

stages of photosynthesis occur in chloroplasts. Both organelles contain their own genetic material. Both have a complex membrane. Mitochondria are powerhouses that produce ATP, but chloroplasts are a combination solar cell and sugar factory.

This is a magnified view of a mitochondrion. It is similar to a chloroplast in shape and size.

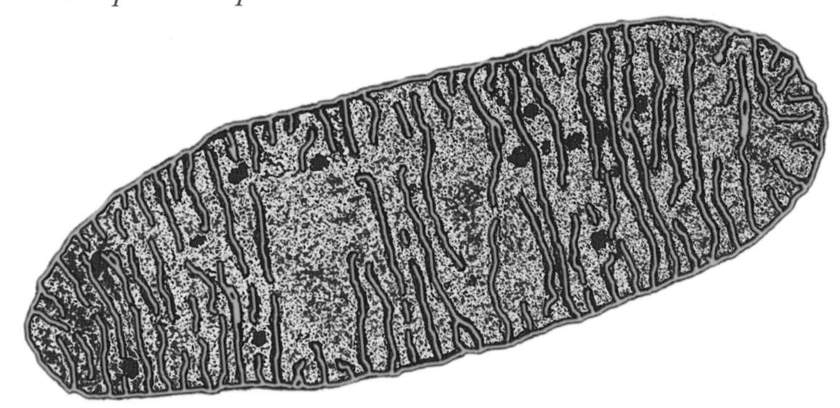

Calvin's C_3 Cycle

The sugar-producing reactions that Melvin Calvin and his coworkers discovered involve nearly two dozen enzymes in a complex series of steps. The most important is the first step, in which CO_2 is linked to a five-carbon sugar called RuBP to form a six-carbon sugar. The newly formed sugar quickly breaks up into two three-carbon compounds. These are the first stable products in the production of carbohydrates. For this reason, the pathway to producing carbohydrates is often called the C_3 cycle.

The first three-carbon compound formed in the Calvin cycle, called 3-PGA, is changed into an energy-rich three-carbon sugar, G3P. This happens in reactions using the energy stored in ATP and electrons from NADPH. Two molecules of G3P then combine to form the six-carbon sugar glucose.

Oddly enough, though, it takes six full turns through the reactions of the Calvin cycle to produce just one glucose molecule. For every six G3Ps that are produced in one turn of the carbon cycle, only one moves on through steps that will create glucose and other carbohydrates. The rest pass through the Calvin cycle again to produce more RuBP and keep the process turning. The three-carbon sugars must go through a number of steps forming five-, six-, and seven-carbon sugars before RuBP is produced. They use up ATP in this process. It may seem inefficient, but the reactions occur quickly and there is plenty of RuBP to keep the process going smoothly.

Scientists call the process of taking the carbon atom from CO_2 and using it in organic materials "fixation." They say that ATP and NADPH provide the chemical energy to "fix" CO_2 to produce carbohydrates.

Earth's No. 1 Protein

The enzyme involved in joining CO_2 and the five-carbon sugar in the first step of the Calvin cycle is a special protein called rubisco. There is more of this protein in the world than any of the millions of other proteins found in living things on our planet!

Glucose, the six-carbon sugar that is produced through photosynthesis, can be used in a number of ways. It may be stored in the chloroplast itself as starch granules. It may be converted into lipids and oils, which are also stored in the chloroplast. Or it may be transported out of the chloroplasts to be used by the rest of the plant cell.

Doing It Their Way

Calvin and his associates learned about photosynthesis by studying green algae. But the cycle of reactions they discovered is very common in the plant world. C_3 plants are found all over the world and include important food crops such as wheat, rice, soybeans, and oats. Not long after the Calvin team discovered the C_3 pathway, two Australian scientists found that some plants use a different pathway to incorporate carbon dioxide into organic materials. The first stable product produced in this photosynthetic pathway is a four-carbon compound sugar instead of the three-carbon one produced in other plants. This has become known as the C_4 pathway. It is common among plants that have adapted to grow in conditions of low moisture and bright sunlight. There are more than one hundred genera of C_4 plants, including corn, sugarcane, and Bermuda crabgrass.

Another Source of Electrons

Oxygen is produced as a by-product of photosynthesis because plants need a source of electrons to replace those that are used in photosynthesis. But water doesn't have to be the source of the electrons. Some photosynthetic organisms, such as sulfur bacteria, use different compounds.

Sulfur bacteria are found in underwater sulfur springs or in volcanoes where a lot of hydrogen sulfide is found. These bacteria use hydrogen sulfide as their source of electrons. Instead of producing oxygen as a by-product, these bacteria produce large amounts of pure sulfur when they grow. It's a good thing that bacteria that produce oxygen evolved or else we wouldn't be here!

When the climate is very dry and the Sun is brightly shining, most plants take measures to conserve water. They close the stomates on the undersides of their leaves to cut down on

A Champion Photosynthesizer

Nearly one-fourth of all the food calories eaten by people and farm animals can be traced to the energy captured from the Sun by corn (maize) plants. Corn is the largest crop in the United States. Each year U.S. farmers plant 70 million acres (28 million hectares) of corn plants—enough to cover the entire state of Arizona. Corn also has another distinction: this C_4 plant is the best photosynthesizer of all the world's major grain crops.

Corn has been around for quite some time. Central American Indians began growing it about seven thousand years ago. By the time Columbus came to the Americas, Native Americans had developed more than two hundred kinds of corn and were growing it from Chile to Canada.

the loss of water into the air. But then CO_2 cannot get into the leaf through the closed stomates. The amount of carbon dioxide inside the leaves drops, and oxygen builds up. The C_3 plant switches from photosynthesis to a different process, called photorespiration, which uses O_2 and produces CO_2 and water but no sugars. C_4 plants have an advantage under these conditions. They can continue photosynthesis at

Pineapples and other CAM plants use a particular type of photosynthesis so they can grow in hot, dry climates.

much lower concentrations of CO_2 because the first product of their photosynthetic reactions, a four-carbon sugar, is used to produce three-carbon sugars and CO_2. The carbon dioxide is then used in the Calvin cycle in nearby cells.

Pineapples, many cacti, and most succulent plants (such as ice plants and jade plants) use a different variation of photosynthesis to thrive in hot, dry climates. They conserve water by opening their stomates only at night. Like C_4 plants, they fix CO_2 into a four-carbon compound. It is stored at night and used in the Calvin cycle during the day. These plants are called CAM plants, for crassulacean acid metabolism, because this kind of photosynthesis was first discovered in the plant family Crassulaceae (jade plants).

In C_4 plants, the C_4 pathway and the Calvin cycle occur in different cells. But in CAM plants, the two cycles occur within the same cell—just at different times. The C_4 pathway takes place at night, and the Calvin cycle occurs during the day.

chapter four

Photosynthesis Today

Living things are joined in our world by a natural balance. All life is linked in a double chain. One chain is made up of energy, and the other chain is the materials of life. Energy flows through life in a one-way stream. Energy comes from the Sun, passes through living things, and eventually leaves our planet as heat, spreading out into space. But the molecules of life—water, oxygen, carbon dioxide, and sugar—circulate in an endless recycled loop.

The Energy Highway

Green plants and other photosynthetic organisms capture energy from sunlight and put it into the chemical bonds of sugar through photosynthesis. Sugar is the universal food and energy source of most living things. Herbivores get sugar from plants, carnivores get it from eating plant eaters, and decomposers get it by breaking down the waste products and dead bodies of the other living creatures.

The food chain is an upside-down pyramid. Most of the energy that a living creature takes in is not stored in its body. It gets used up in metabolism—the chemical processes that go on inside an organism to build up and break down materials. So it takes a lot more energy to make "food" than we get when we eat that food. The higher up a food is on the food pyramid, the more energy it took to make that food. Cattle, which are midway up on the food pyramid, must eat more than 10 pounds (about 5 kg) of grain or an even larger amount of grass to produce just 1 pound (about 0.5 kg) of beef. It is a much more efficient use of energy to eat from the bottom of the food pyramid than closer to the top. It has been estimated that if Americans cut down the amount of meat they eat by just 10 percent, 100 million people could be fed with crops grown on the land freed from growing livestock feed.

The Recycling of Matter

In addition to energy, living things need materials to build their bodies. All the materials that make up living things are part of a circular flow between living creatures. To grow and repair body tissues, organisms take in materials from their environment. These materials are returned to the environment

Three Ways Photosynthesis Keeps Our World Alive

Photosynthesis helps in three ways to keep humans and other animals alive:

1. It provides the fuel from which we get energy.
2. It provides the oxygen to burn the fuel.
3. It keeps Earth's climate in balance.

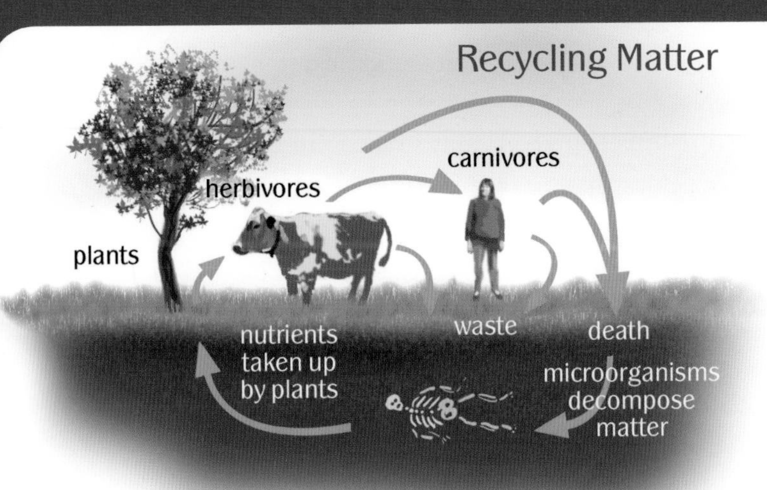

Nothing goes to waste in nature's recycling processes. The nutrients plants pick up from the soil and air are passed on to animals that eat them. Waste products are broken down by bacteria, worms, and other soil dwellers into simpler materials that return to the atmosphere.

through waste products or when the organism dies. Then they are used by other living things.

When a cow eats grass, the waste products are returned to the soil. The matter that was in the original grass may pass through bacteria and earthworms to grass, which then may even be eaten by the same cow.

Unlike other animals, humans take materials from the environment and make them into things that can't be readily used for food, even by decay bacteria. Eventually everything breaks down, but

Did You Know?

The oxygen you are breathing in may once have been breathed in by a dinosaur millions of years ago!

A forklift moves bails of plastic bottles (left) *at the San Francisco Recycling Center. Most bottled water is consumed away from home where recycling isn't an option. For this reason, about 40 million bottles a day go into the trash. It is estimated that 1 million cell phones are retired each week by users in North America. The phone refurbishing company shown here* (right) *processes 10,000 phones a day. The phones are either repaired or sent to recyclers where reusable materials are recovered.*

many man-made materials, such as plastics, take far longer than most natural materials to be turned back into usable forms.

Oxygen—Carbon Dioxide Balance

Animals breathe out carbon dioxide as a waste product during respiration when they burn food to produce energy. Plants use that carbon dioxide to make sugar during photosynthesis and produce oxygen as a waste product. In the big picture of the

Experiment: Observing Carbon Dioxide

Carbon dioxide is an invisible and odorless gas. Limewater (a solution of calcium carbonate—chalk—in water) can be used to test for it.

Put 20 milliliters (.68 ounces) of limewater into a small jar or beaker. Blow into the jar with a straw. The limewater will turn a milky color because there is carbon dioxide in your breath when you exhale.

Put 20 milliliters (.68 ounces) of limewater into a small beaker, and place a candle next to it. Carefully light the candle, and then place a gallon jar over the candle and beaker. The candle burns up all the oxygen and produces carbon dioxide. Will the limewater turn cloudy if you don't light a candle under the jar?

ecosystem, this give-and-take forms a circular flow of materials in a perfect balance.

Plants provide a constant supply of oxygen and use up the carbon dioxide waste products that animals produce to get energy from food. Without photosynthesis, all the oxygen in our atmosphere would be used up and there would be much too much carbon dioxide. The balance between plant photosynthesis and animal respiration keeps the right proportions of oxygen and carbon dioxide in our atmosphere. There is enough oxygen for animals to get energy from food and enough carbon dioxide in the air for plants to make food using the energy from the Sun.

The Oxygen Cycle

Oxygen makes up 21 percent of the volume of our atmosphere. But most of the oxygen on our planet is combined with other elements, such as in water (H_2O), or in rocks, where it is combined with minerals such as silicon, aluminum, iron, calcium, and magnesium. Silicon dioxide is the most common compound in Earth's crust. (This is the basic

Both animals and plants need to take in oxygen, but the amount of this gas in the atmosphere stays the same because of nature's recycling processes. Photosynthesis replaces the oxygen that living things use up.

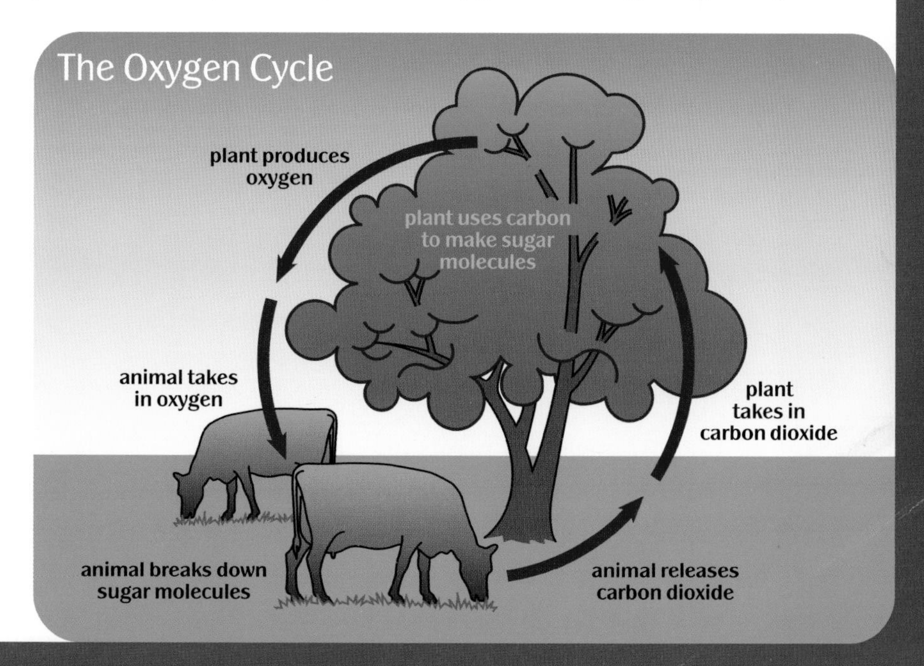

material of sand.)

Oxygen is constantly being removed from the atmosphere in the process of weathering, when the oxygen combines with mineral compounds. About 5 percent of Earth's crust is iron, for example. Almost all iron is combined with oxygen, but half is in a form that could hold more oxygen (ferrous iron). When it is exposed to air, it reacts with oxygen (becomes oxidized), and oxygen is removed from the atmosphere. Scientists believe that this process has locked up an amount of oxygen in the rocks of Earth equal to 70 percent of the oxygen in the atmosphere!

Environmentalists have called tropical rain forests the "lungs" of the planet. They are responsible for 20 percent of the photosynthesis that occurs.

Animals and plants use up oxygen through respiration. It is also used up in the burning of fuels such as oil, gas, and coal. Photosynthesis by green plants and algae makes up the only major source of oxygen that replaces the supply.

Tropical rain forests cover only 3 percent of Earth's surface. But scientists estimate they are responsible for 20 percent of the photosynthesis that occurs on our planet. That's why environmentalists have called them the "lungs" of the planet. However, most experts say that rain forests make very little overall contribution to the amount of oxygen that is added to our atmosphere. Why? Plants undergo cellular respiration just like animals, so they take in oxygen and release carbon dioxide into the atmosphere. Moreover, their fruits are eaten by animals. The CO_2 that was removed from the atmosphere is returned as the animals use the stored food energy. The same happens when the trees die.

The Carbon Cycle

Living plants absorb carbon dioxide and use it to form leaves, stems, and fruit. Carbon dioxide dissolved in water is used by water plants as they grow.

Animals on land and in the ocean eat plants and use the carbon to form shells, skin, bone, and muscle. Animals exhale carbon dioxide when food is broken down. When animals die, the carbon in their bodies is eventually oxidized to carbon dioxide and returned to the atmosphere, except for the carbon that becomes trapped in minerals in Earth's crust.

Large amounts of carbon are locked in rocks and solid materials on Earth. Skeletons of the shells of animals that lived in the oceans millions of years ago can be found in rocks. In Great Britain, the white cliffs of Dover are made up of the

The white cliffs of Dover, England, are made up of the tiny skeletons of sea animals that lived long ago.

Coral reefs are made up of coral skeletons and provide food and shelter to many fish and invertebrates, or animals without spinal columns.

tiny skeletons of sea animals that lived long ago. Coral reefs and some entire islands in the Pacific Ocean are made up of coral skeletons.

Coal, which people dig up out of the ground, and oil, which they pump out of underground deposits, were formed millions of years ago from plants and animals that had died. Scientists believe there is forty times as much carbon in coal and oil deposits as in all the plants and animals now alive on our planet.

The Greenhouse Effect

As the human population grows around the world, people are burning more and more fuels—coal, gas, oil, and wood—to heat homes and businesses and run cars and trucks. Like respiration, burning fuel releases carbon dioxide. Meanwhile, each year people cut down forests to clear land for growing crops and raising animals, for firewood and lumber, and to build houses and factories. The forests of the world produce enormous amounts of oxygen and take in huge amounts of carbon dioxide. Many scientists worry about the disappearing forests. They fear that the amount of carbon dioxide in the atmosphere is increasing at an alarming rate.

Aerial images of clear-cut forests show the large area of trees that are removed for the use of wood and land.

Carbon dioxide gas and other gases in the atmosphere act as an insulating blanket to keep heat from escaping. Without these gases to trap heat, Earth would be too cold for us to live. The heat-trapping effect of the atmospheric gases is similar to what happens in a greenhouse, where the glass windows and roof hold in heat from sunlight. So the effects of carbon dioxide on our planet are often referred to as the greenhouse effect.

Although a certain amount of warming has been good for life on Earth, too much carbon dioxide

Trouble in Grassland

Scientists had hoped that acid rain, which contains nitrogen compounds and other substances that stimulate plant growth, would help to soak up the extra carbon dioxide that our industrial world is producing. But ecologists who studied plots in a Minnesota prairie discovered that this "solution" may not work.

The native grasses in North America are mainly C_4 plants, which not only thrive in hot weather but can get along with very limited amounts of nitrogen. When European settlers came to America, they brought various plants and seeds, which turned into

would trap too much heat. Human activity has poured tons of carbon dioxide into the atmosphere, and Earth's temperature is rising. The ice caps at the North and South Poles have already started to melt. When glaciers melt, the water released causes the sea level to rise. During the twentieth century, the greenhouse effect caused the sea level around the world to rise 6 inches (15 centimeters). This might not seem like a lot. But many cities are at sea level, and beaches all around the world are shrinking as the sea level rises. Scientists worry that if this continues, cities on the coasts around the world may be underwater.

weeds in their new home. The imported weeds are mainly C_3 plants, which are real nitrogen hogs.

When the researchers added nitrogen to the prairie grasslands, the C_3 weeds overran the plots, choking out the native C_4 grasses. The weeds stored a lot of nitrogen, and when they died, they sparked a growth spurt among the soil microbes that break down plant matter. The multiplying microbes allowed the nitrogen that was stored in the weeds to run off and pollute the water, while the plants' carbon was released into the air. As a result, the weeds stored less than half as much carbon as the native grasses, which decompose more slowly.

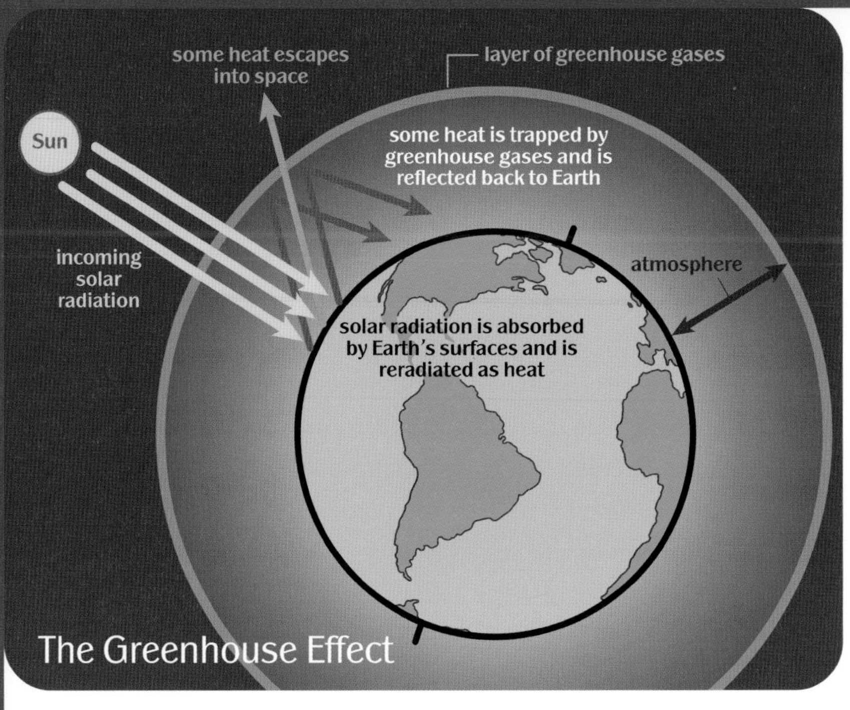

Gases in the atmosphere, such as carbon dioxide, trap and hold some of the heat from sunlight, like the glass walls of a greenhouse. In recent decades, human activities have produced more of these greenhouse gases and Earth's average temperature has been rising.

A furious argument has been raging about whether we should protect our few remaining old-growth forests (such as the giant sequoias in California) or harvest the trees for lumber. Both sides are using the greenhouse effect to support their position. Those who believe we should protect old-growth forests argue that these huge photosynthesizing trees remove large amounts of carbon dioxide from the atmosphere.

Those who want to continue harvesting these ancient trees say that if we replace the old trees with new seedlings, more photosynthesis will occur and more CO_2 will be removed. Young trees do photosynthesize much more, relative to their size and weight, because they grow rapidly. However, only half of a harvested tree is used for lumber. The roots and most of the branches are left behind to decompose. In addition, most of the wood that is harvested is turned into products such as paper, sawdust, and fuel. They are burned or decompose in only a few years. This means that the carbon that was in the tree is put back into the atmosphere as CO_2.

Humans Use Products of Photosynthesis

Many of the basic materials that people use are directly or indirectly dependent on photosynthesis. Textile fibers, lumber, pulp products, vegetable fats, gums, and resins all exist because of photosynthesis. Coal was formed from plants. Natural gas and petroleum were formed by chemical changes in the remains of plants and animals that fed on plants.

Stored Food

Usually plants produce more sugar than they can use right away. Some of the sugar may be turned into starch for storage. Food is stored by plants to use when growing conditions are more difficult. It is also stored to provide energy and growing materials for the next generation of plants.

Sugar and starch can be stored in the plant's flowers, fruits, seeds, roots, and stem. These are the parts of the plant that animals and people usually eat for food.

Experiment: A Starch Test for Photosynthesis

A test for starch can be used to see whether photosynthesis is occurring in a leaf. Plants quickly turn glucose into starch as a temporary storage form when a lot of glucose is being produced. If a leaf is kept in the dark for several hours and then the chlorophyll is removed by soaking the leaf in alcohol, staining the leaf with iodine (to test for starch) will not show much starch present. (The leaf will stain brown.) But a leaf that has been absorbing sunlight all day will stain blue or black when tested the same way. This shows that starch is present.

Carbohydrates

Glucose, also called dextrose, is a simple sugar. It is the main product of photosynthesis. Another simple six-carbon plant sugar is fructose (fruit sugar). Some plants join a glucose and a fructose unit together to form sucrose—the sugar we use to sweeten foods and drinks.

Other important carbohydrates that are made by plants include starch and cellulose. Cellulose is not a food for plants but a building material. Other carbohydrates produced are forms of stored food. In some plants, such as sugar beet, sugarcane, and onion, sucrose is stored as food. But in most plants, starch is

Where's the Food?

Cereal plants and legumes store food in seeds. Carrots, beets, and turnips store food in their roots. Potatoes store food in tubers, which are actually swollen stems even though they form underground. Bananas store sugar and starch in fruits. Artichokes store food in their flowers.

the main storage product.

Glucose is transported from the leaf, where it is produced, to other parts of the plant. But when the Sun is shining, glucose may be produced in the leaf faster than it is used or transported. This extra glucose may be changed into starch as a temporary storage product in the chloroplasts. Later, this starch will be reconverted to sugar and transported from the leaf to other parts of the plant.

At night more glucose is transported from the leaf. The leaves lose weight during the night as the starch is converted to sugar and removed.

In plants such as sugar beets, sucrose is stored as food.

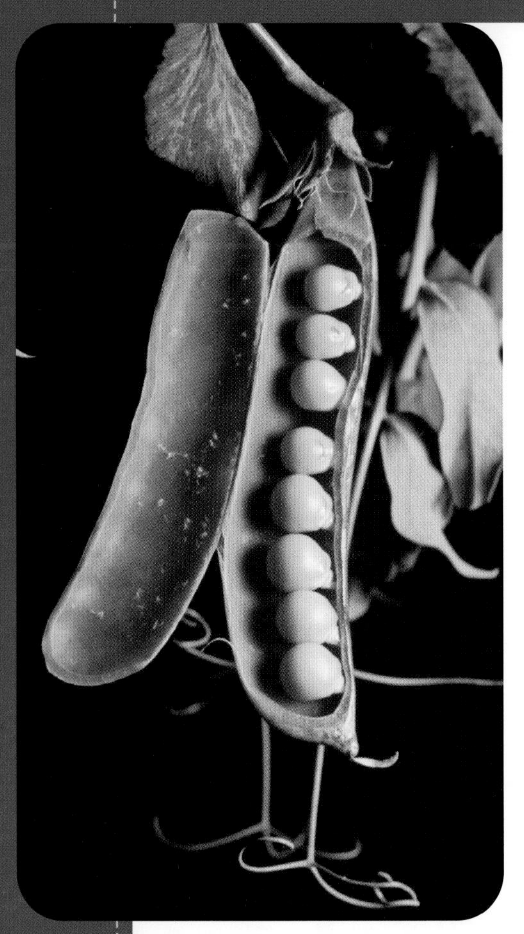

Protein is stored inside the seeds of legumes, such as garden peas.

Proteins

Like carbohydrates, proteins contain the elements carbon, hydrogen, and oxygen, but they also contain nitrogen (and sometimes sulfur) as well. Proteins are stored in seeds (especially those of legumes), corms, bulbs, rhizomes, buds, and storage roots. Large quantities of proteins are needed during growth. When a seed sprouts, stored protein is converted into soluble forms and is transported to the growing seedling.

Amino acids, the building blocks of proteins, are needed by plants and animals. Green plants can produce all the amino acids they need. Animals can form some amino acids from the food in their diet, but most must be obtained from plants. Essential amino acids are ones that animals cannot produce themselves. Nonessential amino acids are ones that animals can transform from other amino acids.

Fats and Oils

Fats and oils are formed from sugars too. They contain the same chemical elements as sugar. However, the ratio of oxygen atoms to carbon atoms is much higher in sugars.

Oils are also an important storage source for energy. They are stored mainly in seeds and also in fruits. Seeds rich in oil include Brazil nut, castor bean, coconut, corn, cotton, nutmeg, peanut, palm, soybean, sunflower, and tung. Fruits high in oil include avocado and olives. Seeds may contain oil as the only stored food, or they may contain both starch and oil. In animals, fats are the main form of food storage. Animals can form many fatty materials from carbohydrates.

What's the Difference Between Fats and Oils?

Chemically, fats and oils belong to the same family. But at room temperature, fats are normally solid and oils are liquid.

Fruits such as olives store a great deal of oil as a source of energy. In fact, 50 percent of the dry weight of olives is oil!

chapter five

Photosynthesis in the Future

Scientists are actively studying photosynthesis. They believe that this process holds important keys to our future. Photosynthesis is tied to almost all human needs. Plants supply all our food, either directly or indirectly, through plant-eating animals. Photosynthesis is also responsible for most of our energy sources—the energy stored in coal, oil, gas, and firewood all came from the Sun through photosynthesis. Photosynthesis also fills many of our needs for fibers and building materials.

If we can learn to understand and control photosynthesis, we can learn how to use our land more efficiently. We can also learn how to increase the supply of food, fiber, fuel, and wood. Once we can effectively reproduce photosynthesis, we can more efficiently use solar energy as a major energy source. Scientists are even using things they have learned about photosynthesis to design supercomputers and medical treatments.

Plant Foods for the Future

As the human population continues to grow, more food will be needed. There will also be less space available to grow that food. Farmers and scientists have already greatly improved the efficiency of growing plant crops for food. Better fertilizers, insect control, and specially bred plants have greatly increased the amount of food that can be grown in smaller areas. They have also improved the nutritional value of many foods.

Scientists have developed even more efficient ways to grow food. Hydroponics is growing food in water. Scientists can add to the water the exact proportion of minerals that will produce the best growing conditions for a particular kind of plant. Dr. William Gericke, a professor at the University of California, coined the term *hydroponics*. In 1930, using hydroponics, he grew a 25-foot (8 m) tomato plant!

This hydroponics tomato tree was set up with artificial lighting in an underground greenhouse. It is intended to grow vegetables as well as train farmers who are interested in the process.

Superfood Producers

Most plants are limited in the amount of food that they can produce. Much of the energy that plants use from the Sun is not used to produce food. Plants must produce roots, flowers, fruits, wood, and other structures that do not contain chlorophyll. The most efficient food-producing plant would have all its parts contain chlorophyll. That way, the whole plant would use the Sun's energy to produce food.

Most algae are microscopic single-cell plantlike creatures. Algae are superefficient photosynthesizers—they can produce a much greater amount of food from the energy they get from the Sun. (Another plus is that most grow in bodies of water and so do not take up any land space.) In Asia and other parts of the world, people eat seaweed, a form of algae. Food scientists are working on various ways to produce good-tasting foods from algae. Adding artificial flavors and aromas and using special processing methods can make "single-cell protein" look, feel, smell, and taste like meat or other common foods. This cheap and plentiful food source can help us keep up with the world's growing needs for food.

Food for Space

If space travel ever gets off the ground, algae could also play a major role. On a long trip, such as a journey from Earth to Mars, it would be difficult to store large amounts of food and oxygen. Algae could

A woman (above) *from a village on the African island of Zanzibar cultivates seaweed on a plantation that she manages. Seaweed farms have become a source of revenue for thousands of women on this island.*

Packages of seaweed soft flour cakes (right) *are sold in a market in Illinois. For reasons such as health, flavor, and availability, Asian food is becoming more popular in the United States and other countries.*

provide a source for both of these needs. Specially bred or "genetically engineered" algae growing in tanks aboard the spacecraft could supply the astronauts with a continuous supply of food. They would also help to maintain the supply of oxygen on board and use up the carbon dioxide that the astronauts breathed out.

Designing Better Plants

Many plants undergo photorespiration. It wastes a lot of the plant's photosynthetic efficiency. Photosynthesis researchers are trying to understand exactly what the function of photorespiration is. When scientists can better understand a process like this, they can change it to design better plants. Researchers hope to produce new crops that will make much better use of the sunlight they absorb.

C_4 plants do not undergo photorespiration. Using genetic engineering, scientists hope to be able to breed some of the features of C_4 photosynthesis into important food-producing C_3 plants. Soybeans, for example, are a great source of protein, but they are limited in the amount of crops they can produce. If soybeans underwent C_4 photosynthesis, scientists could greatly increase the amount of these highly nutritious crops produced in the same space of land.

Photosynthetic Products and Energy

Ethanol, or grain alcohol, is made by fermenting

Soybean crops could be greatly increased if scientists could breed them to develop some of the more efficient features of C_4 photosynthesis.

Career Watch: Photosynthesis

Farming: breeding and growing new, more efficient crops; studying and applying the best conditions for growth; growing single-cell protein in the sea and in factory vats; food science studies to improve taste and nutrition

"Pharming": genetic engineering to transfer genes from one species to another—even animal genes to plant cells—to mass-produce valuable biological products cheaply

sugars and starches. In some places, such as Brazil, it is a major automobile fuel. In some parts of the United States, ethanol is added to gasoline to help reduce the amount of harmful pollutants that are put into the atmosphere with normal gasoline. Ethanol is also easily converted to ethylene, which is used in the petrochemical industry. Scientists hope to be able someday to use cellulose as a cheap source of ethanol. Cellulose is the major component of wood and other plant materials. Some microorganisms are able to convert cellulose to sugar and then to ethanol.

Scientists are also exploring ways to use photosynthesis directly as an energy source. The overall

Solar panels can be used to capture and store solar energy. Scientists are experimenting with similar systems that may provide energy for artificial photosynthesis.

process of photosynthesis is not very efficient. The first steps in the process, in which sunlight is converted into chemical energy, are very efficient. Researchers are trying to build artificial devices that work like plants to capture and harvest solar energy. Scientists have already successfully developed artificial photosynthetic reaction centers that are nearly as good at storing solar energy as chemical or electrical energy as those found in plants. In the future, researchers hope to develop even more efficient solar energy harvesting devices based on the natural process in plants.

When we burn petroleum fuels, other by-products such as hydrocarbons and nitrogen oxides are released into the atmosphere in addition to carbon dioxide. These pollutants are not produced with the artificial photosynthetic reaction centers. Perhaps someday this technology, which is based on the natural process of photosynthesis, will provide us with a clean energy source.

Photosynthesis and Safer Pesticides

Several major herbicides work by interfering with the photosynthetic process in plants. By understanding photosynthesis better, researchers hope to be able to design safer, more effective, more selective herbicides to control weeds.

Photosynthetic Computers?

Computers have become an important part of our everyday lives. Scientists are exploring the idea of using photosynthetic principles to develop smaller, more efficient computers. Plants absorb light, move it to reaction centers, and convert light energy to electrical and chemical energy in a space small enough to fit inside a chloroplast. By learning how plants do this, scientists hope to learn to design computers just as tiny, with working parts only a few molecules big. Researchers have already made some progress in this direction.

Photosynthesis and Medicine

Light has a lot of energy. Sometimes this energy can cause damage to living organisms. Light energy causes aging of the skin and skin cancer in humans and other animals. Plants have developed protective mechanisms that limit the amount of damage light can cause. Scientists are studying how light energy causes tissue damage and are trying to learn from plants ways to prevent light damage from occurring in humans.

Scientists are also using substances related to chlorophyll to fight cancer. These substances will collect in cancerous tumor tissue. When light is shone on the tumors, the cancer cells are killed because of photochemical damage, while the surrounding tissue remains unharmed. In other experiments, the chlorophyll-like substances act as a dye to show doctors the border between cancerous and healthy tissue.

Photosynthesis is an amazing process. Without it, life as we know it could not exist. As scientists learn more about how it works, they are improving our lives in many different ways.

Glossary

antenna pigments: chlorophyll and other pigment molecules in a photosystem that gather light energy

ATP (adenosine triphosphate): a chemical compound used by living organisms for the temporary storage of energy

bacteriochlorophyll: a form of chlorophyll produced by purple bacteria

Calvin cycle: the dark reactions of photosynthesis, in which simple sugars are formed from carbon dioxide

carbohydrates: a family of chemical compounds containing carbon, hydrogen, and oxygen. They include sugars and starches.

carbon dioxide: a gas found in the atmosphere. Its molecules each contain one atom of carbon and two atoms of oxygen.

carnivores: animals that eat other animals

carotenoids: red, orange, and yellow pigments that help in photosynthesis

catalyst: a chemical that helps chemical reactions to take place but itself remains unchanged

cellulose: a fibrous substance in plant cells that humans cannot digest

C_4 plant: a plant in which the first stable product of photosynthesis is a sugar containing four carbon atoms per molecule. Examples include corn, sugarcane, and crabgrass.

chemosynthesis: food production by living organisms using energy stored in chemical compounds

chlorobium chlorophyll: a form of chlorophyll produced by green sulfur bacteria

chlorophyll: a green pigment in plants and certain bacteria that absorbs sunlight energy

chloroplasts: structures in plant cells that contain chlorophyll; the site of photosynthesis

consumers: organisms that eat other organisms in the community

crassulacean acid metabolism (CAM) plants: plants adapted to hot, dry conditions that open their stomates only at night and produce a four-carbon sugar as the first stable product of photosynthesis. Examples include pineapples, cacti, and jade plants.

C_3 plant: a plant in which the first stable product of photosynthesis (via the Calvin cycle) is a sugar containing three carbon atoms per molecule

dark reactions: the chemical reactions of photosynthesis that do not require light energy (also called the Calvin cycle). In these reactions, energy stored in ATP is used to build simple sugars from carbon dioxide.

decomposers: organisms that break down wastes and dead matter, returning their components to the environment

ecosystem: a biological community in which organisms live and interact with one another and their environment

electron: a tiny particle with a negative electric charge. Normally found on the outer part of an atom, electrons can be transferred from one atom to another. Moving electrons produce an electric current.

electron transport chain: a series of chlorophyll and enzyme molecules that transfer electrons in the photosynthetic reactions producing ATP

enzyme: a catalyst (usually a protein) that helps chemical reactions to take place in living organisms

food chain: a sequence of organisms. Each is eaten by the next member in the chain.

food web: the interconnected food chains in an ecosystem

glucose: a simple sugar containing six carbon atoms, twelve hydrogen atoms, and six oxygen atoms per molecule

greenhouse effect: a general warming believed to be occurring because of the heat-trapping effect of carbon dioxide and other gases in the atmosphere

herbicide: a chemical used to kill off undesirable plants (weeds)

herbivores: animals that eat mainly plants

hydroponics: a technique of growing food plants in water, to which mineral nutrients are added

light reactions: the chemical reactions of photosynthesis that take place only in the light. In these reactions, sunlight energy is used to form ATP.

metabolism: the chemical process that goes on inside an organism to build up and break down materials

mitochondria: structures inside a cell in which respiration takes place. Sugars are broken down, forming ATP.

molecule: the smallest unit of a chemical compound that still has all the characteristics of that compound

NADP (nicotinamide adenine dinucleotide phosphate): a compound that accepts electrons and hydrogen ions in the light reactions of photosynthesis, forming NADPH

omnivores: animals that eat both plants and animals

organelles: structures inside a cell that perform a specific function, such as chloroplasts (photosynthesis) and mitochondria (respiration)

"pharming": using genetic engineering techniques to mass-produce valuable biological products cheaply in plants. The name comes from "pharmaceutical" (drug) and "farming."

phloem: fibrous tubes that carry food from leaves to other parts of the plant

photochemical reactions: reactions in which light energy is used to form chemical bonds

photon: a traveling particle of light; a light unit; a packet of light energy

photorespiration: a water-conserving process used by C_3 plants that uses oxygen and produces carbon dioxide but does not form sugars

photosynthesis: a process in which living organisms use sunlight energy to make carbohydrates from carbon dioxide and water, producing oxygen as a by-product

photosystem: groups of 250 to 400 chlorophyll molecules that work together to gather sunlight energy and store some of it in chemical form in ATP

phycobilins: red and blue pigments that help in photosynthesis

producers: organisms that make their own food and provide food for other organisms in the ecosystem

purple sulfur bacteria: bacteria that produce carbohydrates by photosynthesis but do not release oxygen

reaction center: the single chlorophyll molecule in a photosystem that can use light energy for photochemical reactions

respiration: a process in living cells in which carbohydrates and other organic carbon compounds are reacted with oxygen, releasing energy and producing carbon dioxide as a waste product

rubisco: the enzyme involved in joining carbon dioxide (CO_2) and the five-carbon sugar RuBP to form a six-carbon sugar in the first step of the Calvin cycle

RuBP (ribulose-1,5-biphosphate): a five-carbon sugar formed in the Calvin cycle

single-cell protein: protein-rich food produced from algae and processed to give it an appealing texture, appearance, taste, and odor

starches: carbohydrates made up of large numbers of sugar units linked together

stomates: openings on the undersides of leaves through which gases can pass to and out of the leaves. These openings can be closed off by guard cells.

stroma: the inner chamber of a chloroplast

sucrose: table sugar. Its molecule consists of one molecule of glucose and one of fructose (another simple sugar), chemically joined together.

3-PGA (3-phosphoglycerate): the first three-carbon compound formed in the Calvin cycle

thylakoids: stacks of disklike membrane sacs inside the stroma of a chloroplast. Chlorophyll is found in the thylakoid membranes.

xylem: fibrous tubes through which water and minerals are transported from plant roots to the leaves

Select Bibliography

Audesirk, Teresa, and Gerald Audesirk. *Biology: Life on Earth*. 4th ed. Upper Saddle River, NJ: Prentice-Hall, 1996.

Brum, Gil, Larry McKane, and Gerry Karp. *Biology: Exploring Life*. 2nd ed. New York: John Wiley & Sons, 1994.

Burnie, David. *How Nature Works*. Pleasantville, NY: The Reader's Digest Association, 1991.

Crofts, Antony. "Lecture 20: Introduction to Photosynthesis." *University of Illinois at Urbana-Champaign*. 1996. http://www.life.uiuc.edu/crofts/bioph354/lect20.html (February 21, 2007).

Curtis, Helena. *Biology*. 5th ed. New York: W. H. Freeman, 1989.

Davis, P. William, Eldra Pearl Solomon, and Linda R. Berg. *The World of Biology*. 2nd ed. Stamford, CT: Thomson Learning, 1995.

Gust, Devons. "Why Study Photosynthesis?" *Arizona State University*. 1996. http://photoscience.la.asu.edu/photosyn/study.html (February 21, 2007).

Hoagland, Mahlon, and Bert Dodson. *The Way Life Works*. New York: Times Books, 1995.

Massachusetts Institute of Technology. "7.01 Hypertextbook: Photosynthesis Directory." *MIT Biology Hypertextbook*. March 30, 1996. http://web.mit.edu/esgbio/www/ps/psdir.html (February 21, 2007).

Milne, Lorus, and Margery Milne. *Living Plants of the World*. New York: Random House, 1967.

Raven, Peter H., and George B. Johnson. *Understanding Biology*. Dubuque, IA: Wm. C. Brown, 1995.

Raven, Peter H., and Ray F. Evert. *Biology of Plants*. 5th ed. New York: Worth, 1995.

Went, Frits W., and the Editors of Life. *The Plants*. New York: Time, 1963.

For Further Information

Books

Dow, Leslie. *Incredible Plants.* Alexandria, VA: Time Life, 1997.

Fridell, Ron. *Genetic Engineering*. Minneapolis: Lerner Publications Company, 2006.

Hopkins, William G. *Photosynthesis and Respiration*. New York: Chelsea House, 2006.

Juettner, Bonnie. *Photosynthesis*. Farmington Hills, MI: KidHaven Press, 2005.

Seiple, Samantha, and Todd Seiple. *Mutants, Clones, and Killer Corn: Unlocking the Secrets of Biotechnology*. Minneapolis: Twenty-First Century Books, 2005.

Silverstein, A., Virginia Silverstein, and Laura Silverstein Nunn. *Global Warming*. Minneapolis: Twenty-First Century Books, 2003.

Tocci, Salvatore. *Oxygen*. Danbury, CT: Children's Press, 2005.

Vogt, Gregory L. *The Atmosphere: Planetary Heat Engine*. Minneapolis: Twenty-First Century Books, 2007.

———. *The Biosphere: Realm of Life*. Minneapolis: Twenty-First Century Books, 2007.

Websites

ASU Center for the Study of Early Events in Photosynthesis: What Is Photosynthesis?
http://photoscience.la.asu.edu/photosyn/default.html. This site offers activities at the Arizona State University center, articles on photosynthesis, the most recent research and updates, and links to other excellent websites.

Autumn Leaf Color: Why Do Leaves Change Color in the Fall?
http://www.sciencemadesimple.com/leaves.html. This website discusses leaves and their colors and includes science projects and other activities.

Global Warming Kids Site
http://www.epa.gov/globalwarming/kids/gw.html. This website explains global warming, climate, weather, the greenhouse effect, and what we can do to make a difference.

Photosynthesis
http://www.emc.maricopa.edu/faculty/farabee/BIOBK/BioBookPS.html. This website has information, diagrams, and photos about photosynthesis.

Photosynthesis: How Do Plants Make Food?
http://www.ktca.org/newtons/9/phytosy.html. The Newton's Apple Teacher's Guide website includes activities.

Plants
http://www.ed.gov/pubs/parents/Science/plants.html. This site has instructions for some simple experiments on what plants need to grow and how they use sunlight to make food.

Plants and Our Environment
http://library.thinkquest.org/3715/index.html. Learn all about plants and photosynthesis, including diagrams and activities at this site.

SeaWiFS: Studying Ocean Color from Space
http://oceancolor.gsfc.nasa.gov/SeaWiFS/LIVING_OCEAN/LIVING_OCEAN.html. NASA satellite surveys reveal the distribution of phytoplankton in the oceans.

Index

Photo Acknowledgments

The images in this book are used with permission of: © Getty Images, pp. 5, 7, 15, 16, 26, 47 (both), 50, 52 (bottom), 53, 63, 65 (both); © Laura Westlund/Independent Picture Service, pp. 6, 9, 49, 56; Courtesy of the National Oceanic and Atmospheric Administration Central Library Photo Collection, p. 10; Agricultural Research Service, USDA, p. 12; © age fotostock/ SuperStock, p. 13; Library of Congress, p. 17 (LC-USZ62-72127); © Wim van Egmond/Visuals Unlimited, pp. 19, 25 (bottom); © Bettmann/CORBIS, p. 22; © Ron Miller, pp. 23, 29, 31, 46; © Brad Mogen/Visuals Unlimited, p. 25 (top); © Dr. Gerald Van Dyke/Visuals Unlimited, p. 27; © Laguna Design/ Photo Researchers, Inc., p. 35; © Dr. Donald Fawcett & Dr. Porter/Visuals Unlimited, p. 38; © Kurt Scholz/SuperStock, p. 43; © Steve Vidler/SuperStock, p. 52 (top); © Wally Eberhart/ Visuals Unlimited, p. 59; © Dr. James Richardson/Visuals Unlimited, p. 60; © Mauritius/SuperStock, p. 61; © Alvis Upitis/ SuperStock, p. 67; © Inga Spence/Visuals Unlimited, p. 68.

Front cover: © Getty Images

About the Authors

Dr. Alvin Silverstein is a former professor of biology and director of the Physician Assistant Program at the College of Staten Island of the City University of New York. Virginia B. Silverstein is a translator of Russian scientific literature.

The Silversteins' collaboration began with a biochemical research project at the University of Pennsylvania. Since then they have produced six children and more than two hundred published books that have received high acclaim for their clear, timely, and authoritative coverage of science and health topics.

Laura Silverstein Nunn, a graduate of Kean College, began helping with the research for her parents' books while she was in high school. Since joining the writing team, she has coauthored more than eighty books.